Geisla Teles Vieira
Vinícius G. Almeida
Edna N. de Moura

Plant species used in areas degraded by mining

Geisla Teles Vieira
Vinícius G. Almeida
Edna N. de Moura

Plant species used in areas degraded by mining

Revegetation and Environmental Restoration

ScienciaScripts

Imprint
Any brand names and product names mentioned in this book are subject to trademark, brand or patent protection and are trademarks or registered trademarks of their respective holders. The use of brand names, product names, common names, trade names, product descriptions etc. even without a particular marking in this work is in no way to be construed to mean that such names may be regarded as unrestricted in respect of trademark and brand protection legislation and could thus be used by anyone.

Cover image: www.ingimage.com

This book is a translation from the original published under ISBN 978-613-9-66632-4.

Publisher:
Sciencia Scripts
is a trademark of
Dodo Books Indian Ocean Ltd. and OmniScriptum S.R.L publishing group

120 High Road, East Finchley, London, N2 9ED, United Kingdom
Str. Armeneasca 28/1, office 1, Chisinau MD-2012, Republic of Moldova, Europe
Printed at: see last page
ISBN: 978-620-8-05252-2

Our family.

ACKNOWLEDGEMENTS

Firstly to God, because we owe him all our gratitude, for giving us the health, strength and patience to face all the difficulties and emerge victorious.

To our parents, mates, siblings and other family members who were there for us at all times, supporting us and especially understanding our absences when we needed to dedicate ourselves a little more to our studies.

We would also like to thank our friends and colleagues at the university for their companionship and encouragement throughout the course.

To our supervisor Geisla Teles Vieira for her guidance, support and trust. We would also like to thank professors Adriano José de Barros and Jeane de Fátima Cunha Brandão, who helped to finalise our work. To Professor Ariete Pontes de Oliveira, for her patient revision, correction and encouragement.

Finally, we would like to thank everyone who played a direct and indirect part in finalising this work and contributed to our education.

"It is everyone's duty to protect and conserve the greatest national heritage, because the nation that destroys its soil destroys itself."

(Franklin Delano Roosevelt)

Summary

Mineral exploration activities are of great importance and contribution to the Brazilian economy. The requirement for PRAD and environmental care, as described in the project's EIA/RIMA processes, during mine closure, are guarantees that the environment will be returned with quality and biodiversity equal to or close to what was previously found at the start of the project. Soil transposition, seed banks, branch transposition, natural and artificial perches, planting seedlings and nucleation are environmental techniques commonly used to restore the environment. Through a bibliographical review of 18 scientific articles, doctoral theses and research on websites and digital scientific platforms, this research seeks to present the plant species and families most commonly used in the recovery of degraded areas in iron ore mining activities. From the articles consulted, it was noted that the species guandu bean, bragatinga, eucalyptus, fat grass, brachiaria, pau brasil, arnica mineira and skunk grass were the most recurrent in terms of recovering areas degraded by mineral exploitation. The recently published articles show that the Asteraceae, *Melastomataceae,* Podaceae and Orchidaceae families have the largest number of representatives in recovery areas after mineral exploitation, due to their ability to adapt to exposed soil or soil with low nutritional resources. Most of the authors also reinforce the importance and high use of guandu beans, fat grass, brazilwood and eucalyptus as key species in the re-establishment of degraded areas, but it is important to develop a rich seed bank and catalogue the environmental species of the regions before the degrading activity, guaranteeing the success of the recovery and maintenance of regional species.

Key words: Recovery of Degraded Areas; Revegetation; Environmental Restoration; Mine Closure; Seed Bank.

SUMMARY

CHAPTER 1

INTRODUCTION

Iron ore mining is one of the main activities responsible for maintaining Brazil's economy, accounting for the country's annual gross domestic product. The term exploitation is defined by the Michaelis online dictionary as the exploration of mostly non-renewable sources for the purpose of economic utilisation. However, historically, since the beginning of the industrial revolution, with the increased use of machinery and the automation of equipment, mineral extraction has become more constant to meet the needs of developing industries and services. Human beings modify the environment through their actions, causing degradation of areas, the extinction of fauna and flora, the poor quality of water resources and atmospheric changes.

Environmental degradation actions are operations that cause direct damage to the original morphology of the environment, jeopardising the quality of natural resources and the quality of life of the beings that inhabit this environment. Today, mining activities are a benchmark in terms of their capacity for degradation, be it through soil exposure, contamination and excessive use of water resources, changes to the cultural aspect of a region or the suppression of green areas and loss of regional biodiversity.

In the early days of mining activities in Brazil, which began in the 18th century, exhausted mines were abandoned without any measures being taken to reduce socio-environmental risks and impacts. Today, under current legislation, Brazilian mining companies are responsible for mitigating the negative impacts of mine closures and rehabilitating mined areas (FOSCINI; RIBEIRO; SALVADOR, 2009).

Brazil's biodiversity is rich in fauna and flora. Brazil's flora is strongly favoured by the country's tropical climate, which allows for the adaptation of plant species with different characteristics and capable of resisting the country's seasonal variations (LORENZI, 2002). The type and composition of the soil, climatic factors and the availability of resources for plant riches to carry out their photosynthesis and nutrient-seeking activities are factors that justify the presence or absence of a particular species in a given region.

The lack of concern about the protection of trees and grasses leads to irreparable damage to the quantitative factors of species and genetic variation. The expansion of agricultural, livestock and urban areas, mining, fires and firewood extraction are all anthropogenic activities responsible for the loss and reduction of plant mass, leading to environmental degradation of areas (SOUSA et al, 2016).

In Brazil, since 1989, according to Decree-Law 97.632, mining companies have been obliged to submit a Degraded Area Recovery Plan (PRAD), a document that recommends the adoption of procedures to establish or re-establish vegetation cover in degraded areas, a practice known as revegetation (MOURA, 2015). However, there are often difficulties in soil and plant management that can jeopardise the success of revegetation. These difficulties include the use of non-native species, the failure of the species to adapt to the characteristics of the soil, the parameters of water requirements, the high cost and regional climatic conditions, as well as planting problems due to the lack of knowledge of the appropriate technique for recovery areas. Based on the need to reclaim areas after mining activities, this question arises: which tree species are currently most suitable for reclaiming areas degraded by mining? What aspects are used when choosing the tree species to be used?

The process of recovering an area through the use of tree species is of social relevance due to the concept of sustainability, defined by anthropic actions aimed at the maintenance and conscious use of natural resources for present and future generations, related to a mentality and attitude that is ecologically correct and viable in the economic sphere, socially just and with cultural diversification (UN, 2012). There is also the concept of plant succession, represented by Figure 01, which is the alteration of the environment by the species themselves, gradually colonising, replacing and consequently increasing plant wealth. Starting the recovery process with the planting of superior grass species to quickly fix nutrients in the soil, then the process of primary, secondary and then climax succession, in which the recovered region is totally stable and has characteristics very similar to the original, but reaching the climax of a region is a slow process that can take centuries for the return and stability of an entire community.One of the great challenges in the afforestation and restoration of degraded environments is effective recovery in a shorter space of time. In environments with a hostile climate, such as the Brazilian semi-arid region, this challenge is much greater (LACERDA; LIRA-FILHO; SANTOS, 2011). The solution depends, when planning, on the appropriate choice of plant species, such as pioneers, early secondary and late secondary, followed by the need to choose the right tree species, analysing their market availability, speed of development and adaptability to the climatic conditions of each region in order to recover the area at the lowest final cost.

Figure 01- Representation of Ecological Succession

Source: Public domain. Available at <culturamix.com>. Accessed on 04/05/2017.

By means of a bibliographical review of scientific articles, doctoral theses and research on websites and digital scientific platforms over the last 10 years, this research seeks to present the best plant species and families used in the recovery of areas degraded by mining activities. The aim is to provide a solid grounding in the basic and specific contents of species management, to identify the most susceptible species in the recovery of areas after iron ore mining, to specify the tree species most frequently used in the recovery of degraded mining areas and to gather solid material on the species used in mineral exploitation areas so that a manual can be drawn up in the future for enterprises located in south-eastern Brazil.

CHAPTER 2

THEORETICAL FRAMEWORK

The theoretical framework for this research proposal is as follows:

2.1 Mining

Silva (2007) states that mining is one of the basic sectors of the Brazilian economy, which contributes on a large scale to the well-being and improvement of the quality of life of present and future generations, being a fundamental activity for the development of a society, as long as it is operated with social responsibility, always bearing in mind the ideals of sustainable development. But even though it generates income and jobs, mining has a considerable environmental impact.

"Mining obviously has a considerable environmental impact. It intensely alters the mined area and the neighbouring areas where waste rock and tailings are deposited" (SILVA, 2007, p. 02).

2.1.1 Brazilian Mining

According to historical accounts, mining activities in Brazil began in the 18th century, identified as the first major period of mineral extraction in Brazil due to the discovery of gold. Subsequently, the foundations of the Brazilian mining sector were consolidated, placing Brazil as the world's first major gold producer (BARRETO, 2001).

After almost a century, the first gold cycle began to decline. The second mineral cycle began in the 20th century, after the end of the Second World War, becoming effective exploitation from 1960 onwards. "Thus, it can be said that a large part of the current mineral park was built recently and, in particular, during the 1970s and 1980s" **(BARRETO, 2001, p.5).**

Brazil occupies a dominant global position as the holder of large global reserves for a diverse range of metallic and non-metallic minerals, around 40 of them, making it one of the six most important mineral countries in the world.

> The reserves of 11 mineral substances held by Brazil in 2000 are very significant: niobium (1st in the world, 90%), tantalite (1st in the world, 45%), kaolin (2nd in the world, 28%), graphite (2nd in the world, 21%), aluminium (3rd in the world, 8%), talc (3rd in the world, 19%), vermiculite (3rd in the world, 8%), tin (4th in the world, 7%), magnesite (4th in the world, 5%), iron (4th in the world, 7%) and manganese (4th in the world, 1%). (BARRETO, 2001, p. 9)

Barreto (2001) discusses the beginning of the country's visionary approach to obtaining raw materials and Brazil's establishment in the mineral extraction market. He also discusses the beginning of the mineral concern, to produce and preserve, which would later give rise to the National Mining Plan.

> Brazil's mining sector was built on a strategic vision of national development, based on a policy and legislation that encouraged it. Concerns about environmental preservation appeared in the 1980s, although some companies began to incorporate them as early as the 1970s. In this sense, there has been an evolution in the equation of the environmental dimension in Brazil, which has been reflected in the mining sector and which can be identified in three major phases: the first, up until the 1960s, was characterised by a fragmented vision, when environmental protection focused only on a few resources, particularly those more closely related to human health, such as the control of drinking water, concern for certain species of flora and fauna and conditions in the working environment; the second, from the 1970s to the 1980s, began by tackling broader issues, such as environmental pollution and the growth of cities, culminating in a holistic view of the environment as a global ecosystem; and the third, from the 1990s onwards, posed the paradigm of sustainable development as the great challenge, i.e. how to equate economic and social development with preserving the planetary ecosystem. (BARRETO, 2001 p. 6)

For Cabral Junior et al (2008), the lack of technical action in mine development and the lack of planning on the part of public authorities have led to mining conflicts with other forms of land use, such as coffee, eucalyptus and forest plantations. In combination with the lack of training in mine management, Cabral Junior et al (2008) point to the difficulties in controlling and the ineffective environmental recovery of mined areas, as well as the undesirable impacts on the environment, such as altered landscapes, deforestation, the outbreak of erosion and silting, noise and vibration emissions, and air and water pollution.

2.1.2 Degraded Area Recovery Plan - PRAD

With the development of environmental responsibility and the preservation of life, the requirement for the mandatory presentation of the PRAD began, with the central objective of guaranteeing public health and safety through the rehabilitation of areas affected by mining.

> The requirement for the mandatory presentation of the PRAD is based on the principle that areas environmentally disturbed by mining activities must be returned to the community or to the landowner in the desirable and appropriate conditions for the return of the original use of the land or in those necessary for the implementation of another future use, provided that it is chosen by consensus between the parties involved and affected by mining. (LIMA; FLORES; COSTA; 2006, p.397)

For the authors, mine closure must meet legal requirements and take environmental, social and economic characteristics into account at the same time. Although the plans may vary in favour of the beneficiation of a particular region.

> Mine closure planning must meet legal requirements while taking into account the specific environmental, economic and social characteristics of a mine and its surroundings, operations and all the support infrastructure that is part of the mining project. Therefore, the content of closure plans varies to take into account the specific local characteristics of each project. (LIMA, FLORES, COSTA, 2006 p.400)

For the closure of a mine to be successful, Lima, Flores and Costa (2006) indicate that all environmental and regional stakeholders must be taken into account, and that the company must also keep dialogue channels open and maintain a relationship with all players.

Brandi (1994) says that most reclamation plans conceive of mine rehabilitation only as a revegetation process. In addition, most PRADs do not present a plan for monitoring and maintaining the reclaimed areas after the mine closes, a major problem to be solved when seeking to reclaim a degraded area.

2.2 Environmental Degradation and Recovery

In order to analyse the feasibility of restoring vegetation to an area degraded by mining activities, one must be aware of the concepts of recovery and degradation.

2.2. 1Degradation

The concept of degradation of an area has been associated with anthropogenic actions that cause negative and adverse damage to the environment (BITAR, 1997).

> Soil degradation occurs when its natural cover is interfered with, either by simply eliminating it or by replacing it with a poorly managed crop. In the first case, the soil is exposed to erosion and the effects of erosive agents are more or less intense, depending on the soil's resistance to erosion. In the second case, soil degradation can be caused both by erosion and by the deterioration of its properties due to improper use and management. The soil, deprived of vegetation cover and the fixing action of roots, exposed to the direct impact of rain or wind, suffers disintegration and removal of its particles. This effect is complemented by surface water run-off or the abrasion of particles carried by the wind (EMBRAPA, apud COSTA et al., 2000, p.847).

Willians **et al** (1990) uses a concept of environmental degradation related to aspects of morphology, hydrology and edaphological concepts affected by mining, considering **that "degradation of an area occurs when vegetation** and fauna are destroyed, removed or expelled; the fertile layer of soil is lost, **removed or buried; and the quality and flow of the water system is altered"** (WILLIANS **et al**, 1990, p.13). Following the definition of the term, the concept of environmental degradation is established, in which **"it occurs when there is a loss of** physical, chemical and biological characteristics and **socio-economic** development is made impossible**".** It follows that of the environmental bodies degraded by mining activities, the soil is most affected by mineral extraction.

Maschio **et al** (1992) talk about the disruption of environmental degradation, discussing the notion of partial or total destruction of ecosystems and the possible reversal of the damage caused, where irreversibility indicates time and unfeasible expenditure in practical terms.

According to the framework of NBR 10703, published in 1989, the term soil degradation is defined as "adverse alteration of soil characteristics in relation to its various possible uses, both those established in planning and potential uses" (ABNT, 2017, p.16). However, another standard, NBR 13030, specific to mining, defines degraded areas **as "areas with varying degrees of alteration of biotic and abiotic factors, caused by mining activities",** maintaining the notion of alteration, but with no connection to land use (ABNT, 2017, p.56). 2017a; 2017b

So for Bitar (1997), the concept of degradation always seems to be associated with the notion of environmental alteration generated by anthropic activities, which, in a community, tends to include the negative effects on land use as a whole of its effective, planned or potential function.

2.2.1 Recovery

The concept of recovery, characterised by the restitution of a degraded ecosystem or wild population to as close as possible to its original composition (Law 9.985/2000), has generally been presented and discussed in terms of its objectives and goals.

According to Bitar (1997) apud Down; Stocks (1977) the term restoration is best suited to the mining context, involving activities aimed at recreating the initial topography and re-establishing the initial conditions of land use, while any other alternative would correspond to recovery.

According to Cairns Júnior (1986), it is impossible to recover an area with the exact characteristics of the area prior to mining activity, considering the following options from a degraded ecosystem and according to the degree of recovery desired.

> Restoration, representing a situation relatively close to the initial conditions of the ecosystem, but at an intermediate level; rehabilitation, representing the achievement of some of the initial conditions and at a higher level than restoration; and the development of alternative ecosystems, representing conditions quite different from the original ecosystem, but at the same level as restoration. He also uses the terms "recovery", "recuperation" and "regeneration", admitting that although the option of neglecting the degraded area may, over time, tend towards stabilisation, it will occur in conditions very far from the original and to a much lower degree than the other possibilities (BITAR, 1999, p. 28).

Cairns Junior (1986) also mentions the terms repair, recovery and regeneration, inferring that area recovery activities aim to achieve regional climax, although the achievement of regional climax occurs in conditions very far from the original and to a much lesser degree than the other possibilities.

For Barth (1989), it is important to break away from the idea of reclamation as an event that takes place at a specific time or in a specific period of time, and it should always be considered as a continuous planning process that begins before mining and ends long after the activities have been closed.

Brazilian legislation states that the aim of reclamation is to "return the degraded site to a form of utilisation, in accordance with a pre-established plan for land use, with the aim of achieving environmental stability" (Decree 97.632/89).

The authors Willians et al (1990) talk about the art of revegetation applicable to mining, in which the recovery of a site **"will be returned to a form and** use in accordance with a pre-established plan for land use. It implies that a stable condition will be obtained in accordance with the environmental, aesthetic and social values of the surrounding area" (WILLIANS et *al,* 1990, p. 13).

2.3 Ecological succession

Succession is the orderly process of changes in the ecosystem, resulting from alterations to the physical environment by the biological community, ending in a persistent ecosystem type - the climax. Miranda apud Begon et al (2009) define succession as a continuous, directional and non-seasonal scheme of colonisation and extinction of species populations in a community.

According to ODUM (1985), succession is defined according to three parameters:

> (i) it is an orderly process of community development, which is why it is directional and predictable; (ii) it is the result of the community's own modification of the physical environment, i.e. succession is controlled by the community, although the physical environment determines the pattern, rate of change and often the limited set of how development should proceed; and (iii) it culminates in ecosystem stability.

The changes during ecological succession are called beings. Creatures are classified into two groups according to their origins:

- Primary beings are those that occur in previously unoccupied, newly formed habitats such as sand dunes, lava fields, eroded rocks or receding glaciers.
- Secondary beings are those that occur on sites previously occupied by a community soon after a disturbance. They can occur in areas such as abandoned agricultural fields.

In 1916, Frederic Clements proposed a descriptive theory of succession, which consists of the following steps:

-Nudation: Succession begins with the occurrence of a disturbance in an area devoid of life.

-Migration : Arrival of propagules in the environment.

-Ecese : Establishment and growth of pioneer plants.

-Competition : Stage in which the establishment of new species causes competition for space, light and nutrients.

-Reaction : As a result of competition, species are replaced from one plant community to another.

-Stabilisation : The community stabilises after the reaction phases, and a climax community develops.

Miranda (2009) considers the challenge of finding a balance between communities undergoing ecological succession and defines the term climax in his work.

> Communities in equilibrium can be destabilised by disturbances in the environment. After this

> period, communities tend to rebuild, albeit slowly, in a sequence of changes in which species compete for space and resources. Changes occur during the development of succession in the ecosystem until it reaches a state of equilibrium, which is usually called a climax. The climax exhibits the most complete form of exploitation of environmental resources and the occupation of all available niches (Miranda, 2009).

Figure 02 exemplifies the steps and evolution of species through the process of ecological succession.

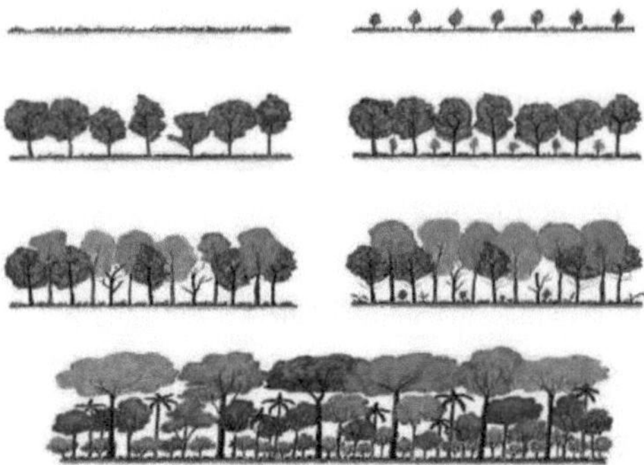

Figure 02- Ecological succession

Source: Handout on Recovering Degraded Areas. Available at <https://goo.gl/6iMr3B>. Accessed on: 10/10/2017.

Preserving and restoring riparian forests within the radius of mining projects are important tools for controlling the quality of liquid effluents and maintaining the regional biota. In adaptive terms, this tool is important for exchanging cultures (native species in the areas to be reclaimed) and ensuring the success of the reclamation.

2.4 Mining Areas Subject to Reclamation

Recovering a degraded area is a process that should begin when the project is being planned and only end once the project's activities have ceased. Thus, the recovery of degraded areas should be understood as a set of actions that are devised and carried out by specialists from different areas of human knowledge, with the aim of re-establishing the conditions of balance and sustainability that existed in the natural system (DIAS; GRIFFITH, 1998).

According to Meira Junior et al (2014), after iron ore extraction, the soil is hardened and impoverished, making natural regeneration or artificial revegetation processes difficult.

> The revegetation process requires the use of appropriate techniques, generally defined according to detailed assessments of local conditions and the use of existing scientific knowledge (PEREIRA et al, 2010). Knowledge and use of ecological and silvicultural information, such as light requirements, plant-animal relationships, frequency and natural density, can lead to knowledge of the ideal combination of species within mosaics of successional stages, similar to those formed in natural tropical forests (MEIRA JUNIOR. apud PEREIRA et al, 2014 p. 284).

According to Brum (2000), reclamation work involving areas degraded by mining activity should cover the following areas: i) mined areas: pits, mining fronts, trenches, galleries in underground mining; ii) solid waste disposal areas, surface soils, tailings, settling basins and sedimentation of processing tailings; and iii).infrastructure areas: these include areas where beneficiation units operate, ore storage and dispatch areas, roads, offices and workshops. As shown in Figure 03, the tailings pile at a mining site in Germany is undergoing a major vegetation recomposition process.

Figure 03- Reclaimed tailings pile Battrop, Germany

Source: Public domain. Available at < revistaecologico.com.br>. Accessed on 10/04/2017.

2.5Soil Characteristics After Mining

The environmental impacts caused by mining are directly related to anthropic action. The extent, magnitude and progress of environmental degradation can be defined according to the exploitation carried out by human activity (KOPENZINSKI, 2000).

Once an area has been used for mining, it becomes a major cause for concern and a source of problems. Depending on the type of terrain and the structure of the pillars in underground mines, there can be problems with subsidence, the lowering of the ground caused by a fracture in the surface, posing a risk to the safety of people travelling through these areas. It is necessary to ensure the stability of structures such as waste rock piles and tailings dams (SÁNCHEZ, 2001).

2. 6Environmental Characteristics that Influence the Growth and Development of Regenerating Species

"Environmental factors such as light, humidity, temperature, soil characteristics and biotic elements affect the growth and development of regenerating plants and the structure of tropical forests" (NAPPO, 2002), since plants have different requirements, "where distribution and abundance are variable along environmental gradients that characterise the site in which they occur" (HAAG, 1985 and SWAINE, 1996).

According to Andrade (1978), temperature and rainfall are the primary and basic factors that determine the establishment of a forest community in a given location. But according to Nappo (2002), with regard to the distribution of different species within the community, the distribution and variety of species is demanded by the availability of light. This is because the diversity of species and the variability of water and light requirements in relation to growth factors favour better use of the light energy that reaches the forest cover.

2.4.1 Light

The effect of light on plant growth depends on its intensity, quality and periodicity (ANDRADE, 1978). Variation in any of these characteristics can affect the development and growth of plants, both quantitatively and qualitatively (FERNANDES, 1998).

> Within a forest, the light radiation that reaches the ground undergoes variations in its quantitative and qualitative characteristics, depending on the species present, the optical properties of the leaves and the density of the successive strata below the canopy (REIFSNYDER ; LULL, 1966, cited by ENGEL, 1989). These changes are of considerable importance in the regeneration and growth processes of a forest ecosystem (NYGREN ; KELLOMAKI, 1984).

The variety and diversity of species, according to Hubbell; Foster (1986), is generally linked to the microhabitats provided by clearings. In time, clearings are, according to Macial (2009), small open areas in the middle of the forest cover and are the result of local disturbance factors, such as the felling of trees or the occurrence of small fires.

According to Nappo (2002) apud Denslow (1987); Denslow; Hartshorn (1994); Martins ; Rodrigues (2002), the great environmental heterogeneity associated with the formation of clearings plays an important role in the biology of tropical forest communities, such as in the establishment, growth and reproduction of many tree and shrub individuals. The formation of a clearing initiates the germination of the seed bank present on the forest floor and drives the growth of seedlings already established in the understorey.

2.6. 2Soil

According to Young (1991), soils have a great influence on the type of plant community present in a locality and, reciprocally, vegetation influences soil properties by removing compounds and supplying organic matter and nutrients.

> The chemical and physical composition of soils is as important to the distribution of plant communities as temperature, rainfall, sunshine, wind speed and other climatic factors. The material of origin influences the establishment of vegetation, with well-drained soils in regions with high rainfall favouring high-density vegetation, while sites with limestone or sandstone as the material of origin support sparse vegetation (NAPPO, 2002 apud EYRE, 1984 and JOSE et al, 1994).

According to Barros (1974), soil depth and texture are the most significant of the physical characteristics in determining the productivity of the site. Nappo (2002) also mentions the influence of texture on forest productivity, which has been reported more as a determining factor in the availability of water for plants.

With regard to the bioavailability of nutrients, soils rich in species are generally poor in nutrients.

> FELFILI (1993), reviewing soil inventories in Brazilian forest typologies with high floristic diversity values, found that, in general, the soils that support these forests are quite poor in nutrients and highly acidic, and that the richness of species is mainly due to the large amount of organic matter on the soil surface, which provides greater moisture retention. KIMMINS; KRUMLIK (1974) observed that a low stock of highly circulating nutrients in the ecosystem has a greater capacity to sustain high diversity than a large stock of poorly circulating nutrients (NAPPO, 2002).

In the studies by Jose et al (1994), conclusive results are presented stating that "high species diversity in tropical forests may be the result of high floristic diversity in microenvironments and/or abundance of physical environmental resources", factors which facilitate the emergence and regeneration of specialised species.

2.7 Methods for Recovering Degraded Areas

Recovering degraded areas requires first defining recovery strategies or methods and choosing the measures to be implemented. The size of the measures must be delimited by the assessment of the degradation carried out previously and by the environmental indicators and parameters used.

Defining the methods requires a careful analysis of the technological fronts available and the likely effectiveness of the measures in correcting or stabilising the degradation. Bitar and Braga (1995) mention three alternatives applied to the recovery of degraded areas, differentiated according to the predominance of the field of scientific knowledge on which they are based: revegetation, remediation and geotechnologies, aimed at the biological, physical and chemical stability of the environment, respectively. However, in practice, the measures are usually applied in combination.

Some specific manuals on the recovery of areas degraded by mining have been produced and edited with contents that illustrate the diversity of possible measures and according to the related application of different techniques, such as the techniques of Bauer (1970), Coppin, Bradshaw (1982), Sendlein et al (1983), Holmberg, Henning (1983), Lyle Jr. (1987), Carcedo, Fernandez (1989) and Alba (1995).

Considering only those measures that aim to ensure the stability of the environment in the short or medium term, recovery methods and techniques tend to be defined according to the main means used to achieve stabilisation: revegetation; geotechnical measures; and remediation. The implementation of the resulting measures may eventually be independent of the type of future land use.

2.7.1 Revegetation

Revegetation methods for areas degraded by mineral exploitation range from the localised establishment of plant species (herbaceous, shrub and tree) to the implementation of extensive reforestation, both for environmental preservation or conservation and for economic purposes, including the generation of favourable conditions for the repopulation of fauna and the regeneration of primitive or original ecosystems.

According to Carcedo et al (1989), revegetation always plays an important role, as it makes it possible to restore the soil's biological production, reduce and control erosion, stabilise unstable terrain, protect water resources and integrate the landscape.

Studies on the feasibility and ideal locations for applying the revegetation technique are currently being discussed and have already been put into practice in Canada, Australia and South Africa, in different types of mines, landscapes and climatic conditions, highlighting the technique for the control and stabilisation of tailings deposits (KENNEDY, 1992).

In Brazil, revegetation techniques have been applied since the late 1970s in large-scale mining operations, as reported in pioneering work on mined areas carried out in bauxite mines in Poços de Caldas, MG, by Alcoa Alumínio (WILLIANS, 1984) and in areas affected by the installation works for the iron ore exploitation project in Serra dos Carajás, PA, by Companhia Vale do Rio Doce - CVRD (FREITAS et al, 1984). Regarding the locations mentioned above, Bitar (1997) says:

> In the case of Poços de Caldas, the revegetation was preceded by terracing, with seedlings of native shrub and tree species planted predominantly along the berms, protected by exotic species (Eucalyptus saligna or grandis) or native species, such as bracatinga (Mimosa scabrela), planted on the crests of the slopes; over time, the exotic species are eliminated to allow the native species to grow.
>
> In Carajás, in the iron mine area, work was carried out using two revegetation techniques: hydroseeding (sprinkling herbaceous seeds in an aqueous medium), aimed at protecting and stabilising around 1,100.000 m2 of cut surfaces and slopes against the erosive action of rainwater, combining fast-growing grass species (Brachiaria decumbes and Melinis minutiflora) with leguminous species (Calopogonium mucunoides and Centrosema pubescens); and manual planting (digging holes, organic fertiliser, planting, crowning and inorganic fertiliser) of around 11.000 seedlings of native and exotic tree species, between September 1983 and March 1984, in the areas degraded by the project's earthmoving activities (borrow areas, the mine's industrial area and the temporary camp area). (BITAR, 1997, p. 38)

Revegetation techniques have also been applied and developed in tailings dam areas, such as experiments in the use of herbaceous species spontaneously established in the mining area, with the aim of controlling erosion processes.

According to Bitar (1997), plant restoration measures have been predominant among the other environmental recovery work carried out by mining companies. The author cites Companhia Vale do Rio Doce - CVRD, where the iron mines in Itabira, MG, the iron and copper mines in Carajás, PA and the gold mines in Teofilândia, BA use the hydroseeding technique in waste rock disposal areas and borrow areas, using grasses and the manual planting of native and fruit species (CAVALCANTI, 1996).

Based on the development of several areas already in the process of recovery through revegetation, especially in large mines in Brazil and abroad, the manual by Willians et al (1990) presents techniques and proposes a sequence of activities for implementing revegetation in areas degraded by mining: planning; drainage works in the area to be mined; removal of the vegetation cover; stripping and opening of the pit (storage of the fertile soil layer and deposition of the tailings); mining and processing; topographical recomposition (filling the pit with tailings, tailings and soil and landscaping); final surface treatment (placement of the fertile soil layer, decompaction and fertility correction); erosion control; revegetation (soil preparation, species selection and planting or seeding); maintenance; monitoring; and future land use.

The significant progress made in revegetation practices for areas degraded by mining in the country in recent years, both in the use of native and exotic species, has been recognised by some authors (GRIFFTH, 1994; GRIFFTH et al, 1996). They note that the approach currently being used is characterised by the quest to improve these practices, but that, as BARTH (1989) observed, this does not guarantee the long-term success of revegetation.

Thus, while recognising that work with exotics may be relevant, Griffth et al (1996) highlight and analyse the results of research into the use of native species and propose an alternative approach, reforming current practices based on what they call an ideal bioeconomic model, carried out in two phases:

> the first, consisting of "rapid growth of vegetation on degraded sites that are also prepared to subsequently receive propagules from the natural communities of the region and facilitate their germination and growth into more evolved vegetative communities"; and the second, with "manipulation of successional dynamics to achieve a self-sustaining and harmonious landscape, in accordance with the land use envisaged in the recovery programme for the area" (GRIFFTH et *al,* op. cit., p.31).

In summary, current case studies suggest a growing tendency to replace current revegetation methods with the use of trees, both in the single planting of exotics, since they do not guarantee revegetation success, and in the planting of natives, since they have been very expensive and time-consuming, or even the mixed process, adopting management practices and inducing natural or spontaneous revegetation.

2.7.2Species Management Plan

Currently, alternatives have been sought for the recovery of degraded areas that allow for a reduction in recovery costs and the return of these areas to an ecological condition closer to the original, i.e. ecological restoration (MARTINS, 2008).

Within this new trend in the recovery of areas, the management and induction of ecological processes have been advocated, with the aim of taking advantage of or stimulating the self-recovery capacity of ecosystems. To this end, knowledge of the communities that colonise degraded areas, as well as the autoecology of the species that make them up, is fundamental for defining restoration methodologies (RODRIGUES; GANDOLFI, 1998).

2. 8Nutrient Requirements

Of the mineral nutrients essential to plants, N, P and S have soil organic matter as their main source. Nitrogen is the most limiting and problematic in production systems, so much so that the use of nitrogen fertilisers has been the main factor in increasing cereal productivity in recent human history (FRANCO; BALIEIRO, 2000).

According to Franco **et al** (2006), in degraded areas, the organic matter content of the soil is very low, "satisfactory plant growth is only possible with the addition of large quantities of organic compost, frequent addition of nitrogen fertilisers or using nitrogen from the air through BNF **(biological nitrogen fixation)".** However, for this to happen it is necessary that

the other nutrients are supplied in a balanced way, and the other factors limiting the efficiency of BNF and plant growth are also taken into account.

"Phosphorus, in addition to being poorly available in most soils, is the main limiting nutrient for BNF and biomass production in natural tropical systems" (PEOPLES; CRASWELL, 1992). On the subject of phosphorus, Siqueira; Franco (1988) apud Franco et al (2006), add

> Its availability is also problematic in the long term, especially in recovery areas, where the soils are generally highly weathered, consisting mostly of Fe and Al oxides and clays. The greater efficiency of P use in these conditions can be achieved by the greater availability of organic matter and through the symbiosis that most plant species in these regions form with mycorrhizal fungi (SIQUEIRA ; FRANCO, 1988; SIQUEIRA, 1996. p. 04).

Most tropical legumes contribute to nitrogen fixation and almost all species tend to associate with mycorrhizal fungi (Siqueira, 1996; Faria ; Campello, 1999). In this way, plant species that form these symbioses are the most suitable for increasing the organic matter content of degraded soils or even productive systems in conditions of low fertility.

2.9 Recovery Techniques

Reclamation techniques aim to rectify the environmental impacts of a given activity and require special solutions adapted to the conditions already established. These solutions, generally used in mining, are based on field observations and technical literature and often involve references to the physical environment.

According to Nolla (1982), in order to implement a recovery plan for degraded soil, it is important to include reforestation with native or exotic species, with the aim of reconstituting the soil, its structure, organic matter and the balance within it.

The tree species must be resistant to the degraded environment and adapted to the region's climate.

Balistieri and Aumond (1997) apud Regensburger (2004) point out that reclaiming areas degraded by mining poses a major dilemma because rapid vegetation cover is needed to protect the degraded site from the intense tropical rains that are often concentrated at a certain time of year. On the other hand, the exclusive use of this technique jeopardises plant succession because rapid cover is only achieved with exotic species.

Franco et al. (2003) further emphasises the importance of species diversity in recovery areas, stating the following

> The sustainability of ecological systems is supported by three pillars: biodiversity, nutrient cycling and energy flow. Thus, to keep the soil productive, any system must include as many plant species as possible in the same crop or in succession, maintain high levels of organic matter along with a high diversity of soil life, and be as efficient as possible in the use of water, light and nutrients (FRANCO et al, 2003).

Measures to recover degraded areas must be planned and implemented according to the different mining activities in which environmental degradation occurs.

The application of the measures is generally based on conventional methods and techniques. The stage of development and application of the measures employed can be identified according to the degree of dissemination in the mines and the effectiveness achieved in correcting or stabilising environmental degradation processes.

2.9.1 Soil transposition

The soil transposition technique is also known as seed bank transposition. It consists of removing portions of the topsoil, along with the litter, from an area at a more advanced stage of succession and placing them in strips or islands in the degraded area. The aim is for these strips or islands to become centres of high species diversity over time, triggering the successional process in the area as a whole.

According to Soares (2008), soil transposition is important because it carries seeds and living beings responsible for nutrient cycling, soil restructuring and fertilisation, which helps to recover the physical and chemical properties of degraded soil and, consequently, to re-vegetate the area.

According to Carrol and Ashton (1965) and Rokich et al (2000), the technique of transposing the soil in conjunction with the seed bank is of great importance.

because it associates the presence of nutrients in the O (organic) horizon of the soil (0-20cm deep) to be transported with the developing seeds in the top layer (0-5cm deep).

> The transposition of the soil seed bank, another important nucleation technique, has been indicated as an alternative ecological restoration technique in degraded areas, due to its low cost and the possibility of containing high floristic richness (CALEGARI et al., 2008; MARTINS, 2009b). It is based on removing the topsoil of the forest soil, which contains the organic horizon (O) and sometimes part of the mineral horizon (A) and even in some cases part of the B horizon (TACEY; GLOSSOP, 1980), which contains high concentrations of organic matter and nutrients, as well as the forest seed bank, and is an important source of seeds of native species (CARROL; ASHTON, 1965; ROKICH et al., 2000).

2.9.2 Seed Bank

The term seed bank is commonly used to designate the current viable reservoir of seeds in a given area of soil. According to Vieira and Reis (2003), the seed bank "corresponds to seeds that have not germinated but are potentially capable of replacing adult plants that have disappeared through natural or unnatural death", and perennial plants that are susceptible to plant diseases and other disturbances.

According to Vieira and Reis (2003) apud Schmitz (1992), the recolonisation of vegetation in a degraded environment occurs mainly through the seed bank in the soil, which plays a fundamental role in the dynamic balance of the area.

> Artificial regeneration in disturbed areas can be better planned if information on the state of the seed bank is effectively collected (TEKLE ; BEKELE, 2000). In this sense, GARWOOD (1989) also emphasises that the richness and abundance of species in the seed bank (associated with seed rain) provide important information on the potential for community regeneration. KAGEYAMA ; GANDARA (2000) emphasise that when choosing a revegetation model, the existence of a seed bank or seedlings of pioneer species and areas with nearby natural vegetation, which can act as a source of non-pioneer seeds, should be taken into account. According to these authors, if there are these two sources of seeds, there is no need to introduce species, and it is possible to use natural regeneration as the most appropriate way of revegetating the area (VIEIRA, REIS, 2003).

The length of time that seeds remain in the bank is determined by physiological factors (germination, dormancy and viability) and environmental factors (humidity, temperature, light, presence of seed predators and pathogens) (GARWOOD, 1989). Understanding the processes of natural regeneration of vegetation communities is important for their successful management (DANIEL; JANKAUSKIS, 1989).

The floristic composition and distribution of the propagules that make up the seed bank are affected both by the types of dispersal of the species present in the area and by those adopted by species in adjacent areas. According to Lopes et al (2010), the seed bank is a dynamic system whose accumulated stock varies according to the balance between inputs and outputs. Inputs come from seed rain, which occurs thanks to dispersal mechanisms. Outflows can occur through physiological responses that are genetically controlled, linked to environmental stimuli (light, temperature, humidity, etc.) or through the death, loss of viability or predation of seeds.

2.9.3 Twig Transposition

According to the definitions of Tres et al (2007), galls are sources of organic matter formed by plant remains (branches, leaves and reproductive material) from the forest. In order to restore an area, this material can be arranged in a disorganised way, forming a complex of plant remains. These branches provide shelter for small animals, as well as maintaining a humid and shaded environment, ideal for the development of plants that are more adapted to this type of environment.

2.9.4 Natural and artificial perches

The use of artificial roosts is recommended for attracting birds and bats, as they provide a roosting area for these animals that can move between forest remnants. Through faeces and material regurgitated by these animals, seeds are deposited in the vicinity of the perches, forming nuclei of diversity.

Natural roosts are obtained by planting fast-growing trees that have a favourable crown for birds and bats to roost in, and may have fruit that attracts these animals. Trees remaining in the area can also be used. Artificial perches can be built with bamboo poles, eucalyptus posts, dead tree stems or recently felled trees (with environmental authorisation). Fast-growing vines can be planted around the perches to help attract birds.

2.9.5Planting seedlings

According to Soares (2009), planting seedlings is an effective way of extending the nucleation process. It can be carried out in different ways, in terms of the arrangement of seedlings in the field. One form of planting would be random, where the seedlings are planted without defined spacing. Another model is line planting with pioneer and non-pioneer species, using a spacing of 2 x 3 m or 2 x 2 m (Figure 4).

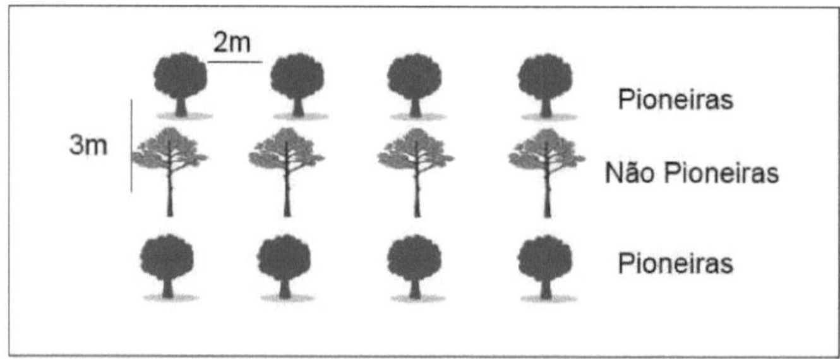

Figure 04- Planting model in alternating rows of pioneer and non-pioneer species.

Source: Soares (2012). Accessed on: 07/09/2017.

There is also the planting of seedlings in dense groups, where the spacing **between seedlings is small. An example of this type of planting is the so-called "Anderson groups", where 3, 5 or 13 seedlings are planted 0.5 metres apart in a** homogeneous or heterogeneous manner, as shown in figure 05.

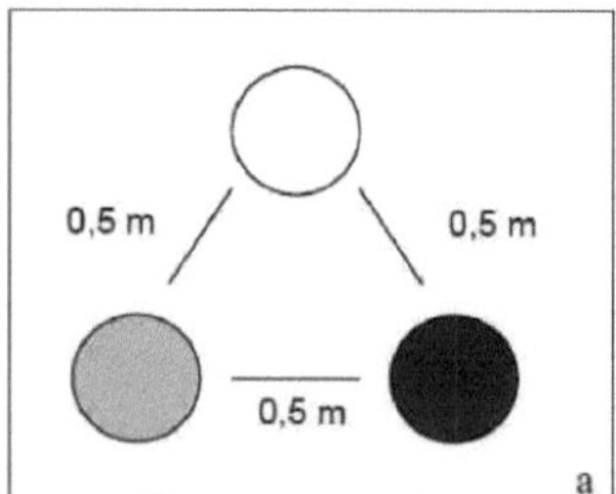

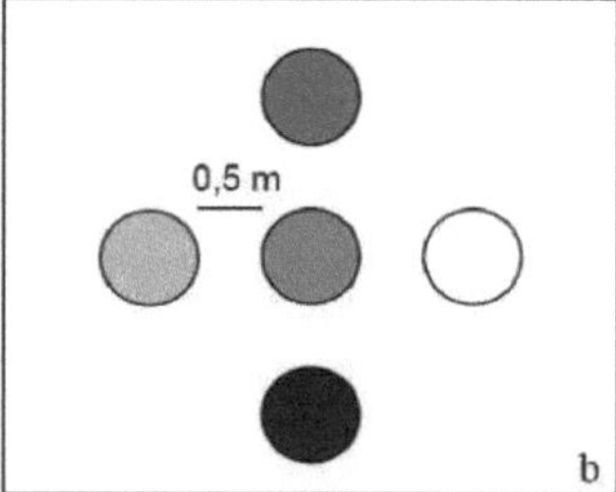

Figure 05- **Models for planting seedlings in "Anderson groups"** (a - with 3 seedlings and b - with 5 seedlings).

Source: Soares, 2012. Available at <ufjf.br/ecologia/files/2009/11/estagio_Silvia_Soares1.pdf>. Accessed on 07/09/2017.

2.9.6 Nucleation

According to Soares (2009), nucleation is the ability of a particular species to significantly improve the environment, facilitating the occupation of that area by other species. Thus, from islands of vegetation or nuclei, secondary vegetation expands over time and accelerates the process of natural succession in the degraded area.

Nucleation techniques aim to form microhabitats in favourable nuclei **to open up a series of "eventualities" for natural regeneration, such as** the arrival of plant species of all life forms and the formation of a network of interactions between organisms.

According to Bechara (2006), **the aim of nucleation is to promote "ecological triggers" by increasing the likelihood of the formation of a range of** successional **events** that could be effective in recovering the degraded area.

The restoration and successful recovery of degraded areas, then, according to Reis et al (2006), "is **characterised by various techniques that are** implemented, never in a total area, but always in nuclei, in order to leave **open** spaces **for the eventual to express itself", a** combination represented by figure 06. Each of the nucleating restoration techniques has different functional effects and particularities which, together, produce a variety of natural flows over the degraded environment, maintaining key processes and helping to restore the complexity of natural system conditions.

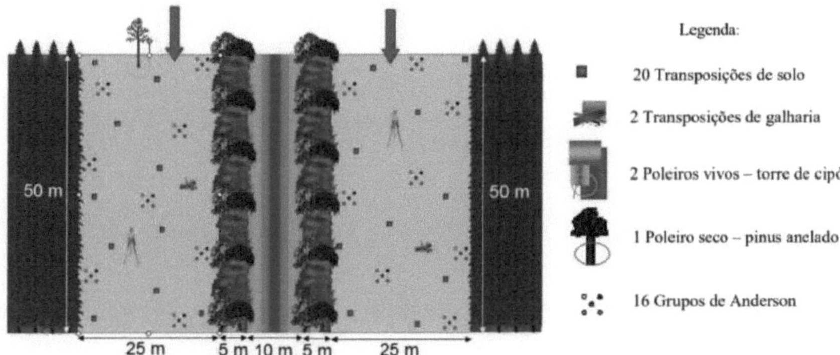

Figure 06: Restoration based on the stochastic successional model, with a diversity of spaces as an opportunity for natural regeneration.

Source: Reis et al, 2006. Available at < https://goo.gl/MqLGM9>. Accessed on: 10/11/2017

Within this new concept of restoration, a set of actions must be proposed in order to nucleate the natural flows that will facilitate nature's recovery. Defining the best strategy for restoring an ecosystem requires understanding a series of processes that occur in the community, taking into account the context in which the environment is inserted.

2.10Costs for an Enterprise

In proposals for expenditure on measures to recover areas degraded by mining, expenditure on mine closure activities involving the restoration of vegetation in the areas is accounted for in the planning and management of environmental control, and it is difficult to measure exact figures for the process.

According to Neto and Petter (2005), environmental economics would be a way of controlling the action of degrading activities, with the application of instruments economic. These instruments would be able to provide economic solutions for environmental management in conjunction with legal instruments.

There are also the costs of planning and managing activities carried out at the enterprise, where the total costs of carrying out environmental measures, such as the recovery of areas, are entirely included in operating costs. Estimates of costs for environmental measures in mining activities are uncommon, but the use of native species can reduce costs as long as the soil is prepared to receive the species.

CHAPTER 3

METHODOLOGY

As this is a research project that seeks to qualify the tree species most commonly used in revegetation of areas after iron ore mining activities, this work is categorised as theoretical and case study research, which aims to generate knowledge about the current species used in the recovery of these areas, involving real cases that are already under development.

As for the research approach, this is qualitative research in the sense of objectifying the phenomenon, seeking to hierarchise the actions of describing, understanding the topic researched and explaining. "Qualitative research tends to emphasise the dynamic, holistic and individual aspects of human experience, in order to grasp the totality in the context of those who are experiencing the phenomenon" (GERHARDT; SILVEIRA, 2009). The search for precision in the relationship between the global and the local in the context moves the research towards a qualitative approach, respecting the interactive nature between the objectives sought, theoretical orientations and empirical data, seeking to define the most reliable results possible.

As far as the nature of the research is concerned, this is basic research with no practical application due to the length of time it takes to acquire data when the revegetation technique is applied. In terms of objectives, the research is considered descriptive, as it aims to "describe the facts and phenomena of a given reality" (TRIVINOS, 1987). Although Trivinos (1987) mentions that descriptive studies can be criticised because there can be an exact description of phenomena and facts, Gerhardt; Silveira (2009) say that descriptive studies avoid the possibility of investigation through the analysis of applied techniques.

In terms of technical procedures, the research is classified as bibliographical research, a case study and documentary research. Bibliographical research is based on a survey of theories that have already been analysed and published in written and electronic media, such as books, scientific articles and websites (FONSECA, 2002). A case study is a study that aims to gain an in-depth understanding of the how and why of a particular situation that is supposed to be unique in many respects, seeking to discover what is most essential and characteristic about it (FONSECA, 2002). Documentary research makes use of more diverse and dispersed sources that have not been analysed, such as newspapers, magazines, official documents, etc.

(FONSECA, 2002). The research is based on other documents, bibliographies and publications of practices that have already been carried out, mainly because it is not an unprecedented piece of research. It is carried out in the majority of iron ore mineral extraction ventures in the final plan of activities, with the aim of closing mines correctly in accordance with current legislation.

In this proposal, this work consisted of searching for articles, doctoral theses, publications in books, conference papers and manuals on the recovery of degraded areas **that dealt with the topic** "selection of tree species used in the recovery of degraded areas in iron ore mining". To do this, a search was made on search platforms using the descriptors: mining, mine closure, recovery of degraded areas, tree species, revegetation, floristic recomposition, published in the last 10 years.

CHAPTER 4

RESULTS AND DISCUSSION

Eighteen scientific papers were found with the descriptors "mining", "mine closure", "recovery of degraded areas", "tree species", "revegetation", "floristic recomposition". During the search for bibliographic material, it was difficult to find articles that covered many of the descriptors. As a result, we chose articles that covered at least two of the descriptors.

The analysis of the 18 articles (which also included other biomes, but dealt with revegetation) revealed that the species guandu bean (Cajanus cajan) (33.33%), bragatinga (Mimosa scabrella) (27.78%), eucalyptus (Eucalyptus saligna) (27.78%), fat grass (Melinis minutiflora) (27,78%), brachiaria (Brachiaria decumbens) (22.23%), pau brasil (Caesalpinia echinata) (22.23%), arnica mineira (Lychnophora pinaster) (16.67%) and skunk grass (Andropogon ingratus) (16.67%) were the most frequent in the publications, indicating their recurrent use in practice. Table 01 shows the most recurrent species in the publications consulted, listing the percentage of citations in the 18 articles consulted, and informing about their nature, whether exotic or native to the Brazilian flora.

Table 01: Most recurrent plant species in publications in relation to the percentage of citations and their nature (exotic or native).

Species	Quantity Quoted	Percentage*	Native or Exotic
Cajanus cajan	6	33,33%	Exotic
Mimosa scabrella	5	27,78%	Native
Eucalyptus saligna	5	27,78%	Exotic
Melinis minutiflora	5	27,78%	Exotic
Brachiaria decumbens	4	22,23%	Exotic
Caesalpinia echinata	4	22,23%	Native
Lychnophora pinaster	3	16,67%	Exotic
Andropogon ingratus	3	16,67%	Exotic
Axonopus marginatus	2	11,11%	Native
Andropogon bicornis	2	11,11%	Native
Phaseolus vulgaris	2	11,11%	Native
Piptadenia gonoacantha	2	11,11%	Native
Anadenanthera peregrina	2	11,11%	Native
Cecropia glaziovii	2	11,11%	Native
Cecropia hololeuca	2	11,11%	Native
Paspalum scalare	2	11,11%	Exotic
Centrosema pubescens	2	11,11%	Native
Mimosa calodendron	2	11,11%	Native

Caption: *Species citation frequency in relation to the 18 articles analysed. Source: Author's own work.

Graph 01 represents the results collected in Table 01 visually, where the "Other species" percentage refers to the species cited in 11.11 per cent of the articles: Axonopus marginatus; Andropogon bicornis; Phaseolus vulgaris; Piptadenia gonoacantha; Anadenanthera peregrina; Cecropia glaziovi; Cecropia hololeuca; Paspalum scalare; Centrosema pubescens; Mimosa calodendron.

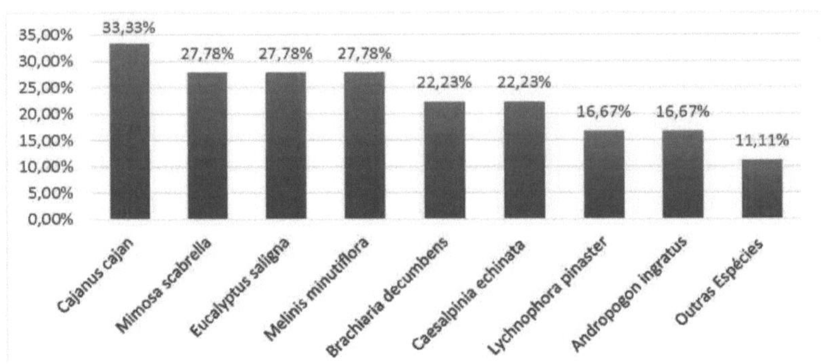

Graph 01: Plant species most cited by authors in the recovery of iron ore mining areas.
Source: Own authorship.

Graph 02 shows a comparison between the use of native and exotic species in the recovery of areas degraded by mineral exploitation, with reference to the results obtained in Table 1. The results showed that more species native to Brazilian biomes (61.10%) were used in the publications analysed. Exotic species can be highly adaptable to environments, showing rapid growth and ground cover. However, they have no natural predators or competitors, which can lead to biological invasion, with the establishment of plant pests, which is why, according to the authors' conclusions, the species are commonly used when native species have difficulties adapting to the environment.

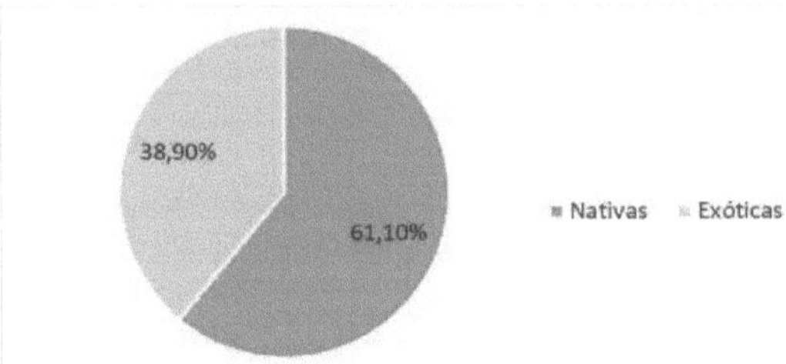

Graph 02: Percentage of native and exotic species used in mineral exploration areas.

Source: Own authorship.

According to Neri et al (2011), the most cautious part of recovery projects is choosing species with the potential to start the plant succession process in a degraded area. The use of native species in plantations for the recovery of degraded areas (RAD) is highly important, as this conserves the region's biodiversity and reduces the costs of seedling production and transport (MOREIRA, 2002 apud SAMPAIO, 2007). Jesus et al (2016) reinforces this by saying that in order to guarantee the success of plant recomposition and achieve environmental restoration of the area degraded by mineral exploitation, it is necessary to combine native and exotic species.

Based on the 18 articles analysed, it can be seen that an average of three different species were used in all of them, and never just one species. The diversity of forest species in plantations to recover degraded areas is extremely important, as it brings various benefits such as the contribution of organic matter to the soil and the redistribution of nutrients (PEREIRA; RODRIGUES, 2012).

After analysing the species described in the articles consulted, graphs 03 and 04 were drawn up based on the families most commonly found and catalogued in areas of recomposition after iron ore mining.

The most frequent families in the survey, relating grass and tree species in quantitative order, were: Fabaceae (or Leguminosae), (17 species), Asteraceae (10 species), Melastomataceae (05 species). It is worth noting that the families defined as "other" are not very expressive individually in the study regions, with this group comprising 27 families. Graph 03 shows the result of analysing the 18 articles consulted, with a representation of the most common families in areas of plant recomposition after mineral exploitation. Some representatives of the most frequent families are shown in Figure 7.

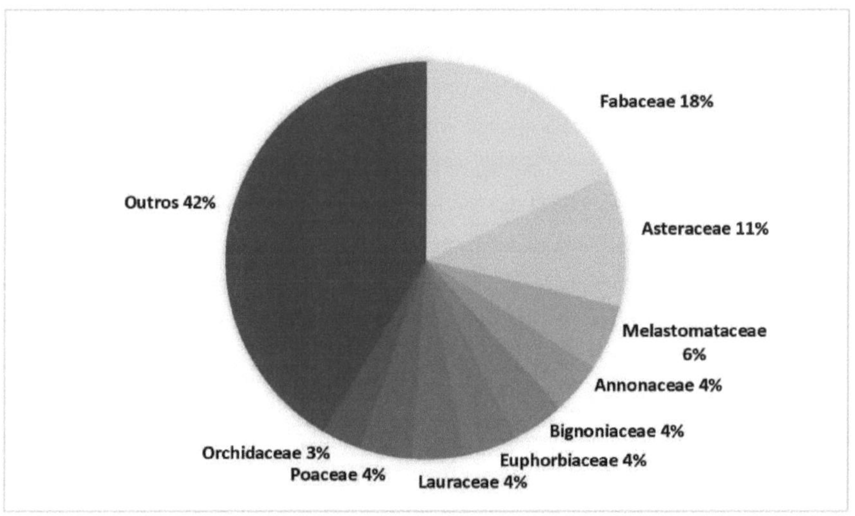

Graph 03 - Most representative botanical families in the recovery of areas degraded by mining

Source: Own authorship.

Figure 7: Specimens from the Fabaceae (A), Asteraceae (B) and Melastomataceae (C)

families.

Source: Public domain. Search tool, family names. Accessed: 28/11/2017.

According to Franco et al (2003), areas after mining activities tend to have exposed, nutrient-poor soil with sandy characteristics and compacted areas, requiring primary correction of the soil's physical and chemical characteristics, as well as systematisation of slopes and the soil that will receive the species, preferably primary succession and leguminous plants, to comply with the PRAD's requirements.

Some larger species, such as the shrubs M. calodendron *(Fabaceae), L. pinaster (Lumbrineridae) and S. glabra (Chloranthaceae),* based on the articles consulted, are often found in association with seedlings of other species, mainly orchids. These species provide shade, accumulation of organic matter and physical protection from the wind, softening environmental conditions and favouring seedling recruitment and establishment.

In terms of quantification and extent of revegetated areas, Eucalyptus sp. is present in the vast majority of the articles consulted. According to Araújo et al (2008), "when exotic species are used in PRADs, they generally have a wide dispersal range, spreading more quickly and overlapping with native species", which **can** form patches in the landscape and prevent the development of native species. In this way, the presence of eucalyptus in all the areas, regardless of the PRAD model adopted, can be explained by the species' ecological behaviour of rapid growth and dispersal and by the location of homogeneous plantations of this species in the adjacent areas.

After reading the 18 articles, iron-removing primary succession plant species were identified: Stachytarpheta glabra, Lychnophora pinaster and Mimosa calodendron. These species are tolerant to high concentrations of heavy metals in the substrate and should therefore be used under these conditions. Species with clonal growth such as grasses (mentioned above), lianas and epilithic orchids are indicated as primary individuals in plant succession.

According to Jacobi et al (2008), "one must also take into account the hierarchical position that the species occupy in the ferruginous grassland community", which supports the recommendation of Andropogon ingratus, Bulbostylis fimbriata, Sophronitis caulescens and Lychnophora pinaster, as well as Sebastiania glandulosa, which had the highest importance values in the community and are frequent in other areas of mineral extraction canga.

After analysing the article by Jacobi, et al (2008), it can be concluded that the Asteraceae family showed greater coverage capacity when compared to other **families in relation to the "vegetation cover" factor in RAD mining. These** data are **analysed** in Graph 04 and Table 02. The species mentioned in this study include *Lychnophora pinaster, Symphyopappus brasiliensis, Baccharis serrulata, Chromolaena sp., Trichogonia sp. and Trixis vauthieri.* The factor (vegetation cover) is considerably important in the recovery of areas degraded by mining because with the change of seasons, the leaves tend to fall and recharge the soil with nutrients and organic matter for its recomposition (JACOBI, et al, 2008). The Poaceae family was relevant in terms of area coverage, as it accounted for 17.53% of the vegetation cover in one area. The usual described species of this family, according to research by Jacobi et al (2008), are *Andropogon ingratus, Paspalum scalare and Axonopus siccus.*

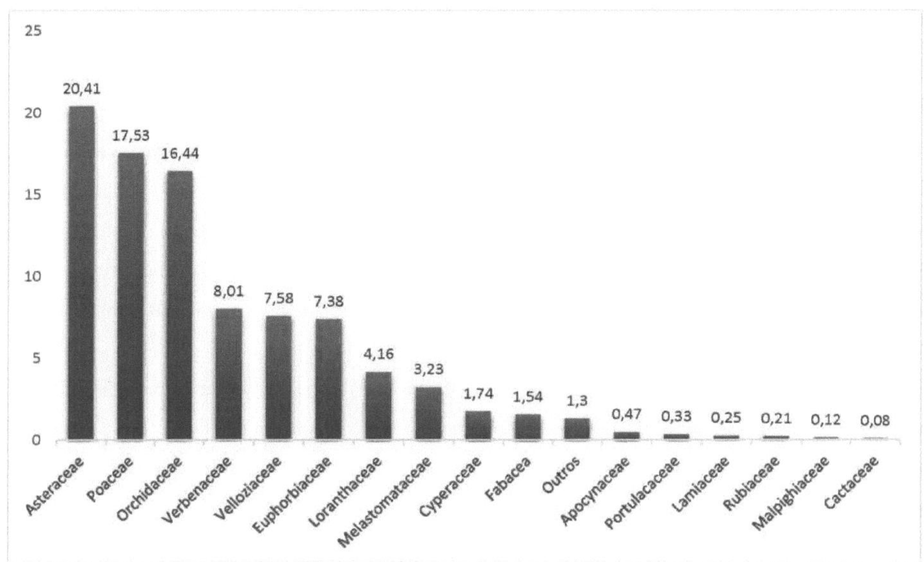

Graph-04: Plant families in terms of vegetation cover in RAD mining.
Source: Own authorship, extracted from Jacobi *et al* (2008).

Table 02: Plant families in terms of vegetation cover in RAD mining.

Species	Family	CVI (%)	Species	Family	CVI (%)
Andropogon ingratus	*Poaceae*	13,67	*Baccharis reticularia*	*Asteraceae*	*0,78*
Lychnophora pinaster	*Asteraceae*	12,35	*Epidendrum secundum*	*Orchidaceae*	*0,68*
Bulbostylis fimbriata	*Cyperaceae*	10,65	*Arthrocereus glaziovii*	*Cactaceae*	0,43
Sophronitis caulescens	*Orchidaceae*	9,99	*Ditassa mucronata*	*Apocynaceae*	*0,42*
Stachytarpheta glabra	*Verbenaceae*	8,01	*Portulaca hirsutíssima*	*Portulacaceae*	0,33
Sebastiania glandulosa	*Euphorbiacea e*	7,33	*Eriope macrostachya*	*Lamiaceae*	*0,25*
Acianthera teres	*Orchidaceae*	5,77	*Axonopus siccus*	*Poaceae*	*0,24*
Symphyopappus brasiliensis	*Asteraceae*	5,53	*Galianthe sp.*	*Rubiaceae*	*0,21*
Struthanthus flexicaulis	*Loranthaceae*	4,16	*Heteropteris sp.*	*Malpighiaceae*	*0,12*
Vellozia compacta	*Velloziaceae*	4,13	*Chromolaena sp.*	*Asteraceae*	*0,12*
Paspalum scalare	*Poaceae*	3,62	*Tibouchina sp.*	*Melastomatacea e*	*0,08*

Vellozia grass	*Velloziaceae*	3,45	*Mediterranean periwinkle*	*Fabaceae*	*0,08*
Tibouchina multiflora	*Melastomataceae*	3,15	*Trichogonia sp.*	*Asteraceae*	*0,05*
Lagenocarpus rigidus	*Cyperaceae*	1,69	*Croton serratoideu*	*Euphorbiaceae*	*0,05*
Mimosa calodendron	*Fabaceae*	1,46	*Trixis vauthieri DC.*	*Asteraceae*	*0,03*
Baccharis serrulata	*Asteraceae*	1,16			

Source: Adapted from the results of Jacobi et al (2008).

The analysis of the 18 articles indicated that legumes, grasses and tree species were frequently used in the revegetation process. We chose to discuss and present data from the literature that justifies the recurrent use of these types of plants in the revegetation process.

Recent articles show that green manure is very important for the rapid recovery of soil properties (LONGO et al, 2010; NOGUEIRA, et al, 2012; AZEVEDO, et al, 2007). According to Longo et al, 2010, the use of black mucuna (Mucuna aterrima - Leguminosae), guandu beans (Cajanus cajan - *Leguminosae) and brachiaria (Brachiaria decubens* - Gramínea), together with organic fertiliser, provided good soil cover, which could positively modify some properties of the soil/substrate in question.

Most leguminous tree species are capable of forming an efficient symbiosis with diazotrophic bacteria (FARIA *et al*, 1999). Due to this association, different species of the family have been used to condition acidic and low-fertility soils, such as tropical soils, because they have low nutritional requirements and tolerance to soil acidity (DIAS *et al*, 1991; FERNÁNDEZ *et al*, 1996; BALIEIRO *et al*, 2001). According to Souza (2009), the *Leguminosae* botanical family is one of the most important in the tropics, with herbaceous, shrubby and arboreal representatives distributed in more than 650 genera and 1800 species, with 1500 species distributed in 175 genera found in Brazil.

The ability to improve soil fertility through the cultivation of leguminous trees was also approved by Gomes *et al* (2007) who, investigating the return of nutrients to the soil through litter, found that in homogeneous stands, *Caesalpinia echinata* (pau-brasil) contributed to the increase in organic matter and macronutrients on the soil surface, especially the amounts of nitrogen and potassium, as did *Caesalpinea ferrea* (pau- ferro), which despite being a tree species, the fall of its leaves contributes to forage and soil nutrition.

The success of recovering a degraded area is closely linked to the species used for recovery. Well-adapted, fast-growing species are essential for the design of vegetation covers, which reduce variations in soil temperature, retain a greater amount of water, reduce evaporation and surface run-off and prevent erosion processes

(NOGUEIRA et al, 2012). In Table 01, Nogueira et al (2012), and Azevedo et al (2007), cite species commonly used in recovery areas, relating the general pattern, applicability of use and disposal and environmental adaptation.

Table 01 - Grass and herbaceous leguminous species used in the recovery of areas degraded by mineral exploitation.

Scientific name	Family	Plant pattern	Use	Environmental adaptation
Arachis pintoi	*Fabaceae*	Herbaceous	Production of phytomass and fixation of microorganisms	Stable soils
Andropogon ingratus Hack	*Poaceae*	Grass	Soil forage	-
Brachiaria decubens	*Poaceae*	Grass	Soil forage	-
Centrosema pubescens	*Leguminosae*	Grass	Soil forage	Various soils, pastures and crops
Cajanus cajan	*Fabaceae*	Grass	Grain, fodder and green manure	-
Crotalaria juncea	*Fabaceae*	Grass	Green manure	Various soils (R.A.D.)
Gliricidia sepium	*Fabaceae*	Herbaceous/Shrub	Fodder and firewood	Low acid and moist soils
Lychnophora pinaster	*Asteraceae*	Shrub	Medicinal and soil fodder	Soils with high ferrous concentrations
Mucuna aterrima	*Leguminosae*	Grass	Fodder and green manure	-
Mimosa scabrella	*Leguminosae / Mimosoideae*	Grass	Firewood, cellulose and fodder	Pioneer species. Acidic soils and reclamation areas.
Neonotonia wightii	*Fabaceae*	Grass	Soil forage	Various soils
Pueraria phaseoloides	*Fabaceae*	Herbaceous	Production of phytomass and fixation of microorganisms	Stable soils
Lolium multiflorum	*Poaceae*	Herbaceous	Green manure	Various soils (R.A.D.)
Glycine wightii	*Phaseoleae*	Herbaceous	Phytomass production	Various soils (R.A.D.)
Melinis minutiflora Beauv	*Poaceae*	Herbaceous	Green manure	Various soils (R.A.D.)

Source: Adapted from Nogueira *et al*, 2012 apud CEMARH (2010); and Azevedo *et al*, (2007).

From the feasibility study and description of species to be planted in mineral exploration areas as a form of restoration, it was noted that although legislation number 6.938 of 31 August 1981 is already widely applied in mineral exploration companies, little information on species and families is made public, which makes it difficult to access more information on plant species in the revegetation process in mining.

Regions after mineral exploitation have the characteristics of rocky outcrops and unprotected soil, which suggests primary ecological succession. Based on **these conditions,** "plant communities on rocky outcrops generally have a low diversity of species compared to the **surrounding** ecosystems" **(RIZZINI, 1997).**

Currently, for new areas to be exploited, it is common to set up a rich seed bank in addition to creating and maintaining Permanent Protection Areas (PPAs) in order to guarantee the planting of the right species and reduce the chances of species not adapting or techniques failing.

In the studies carried out by Nappo (2002), which aimed to describe the dynamics of natural regeneration in a mineral exploitation area in Poços de Caldas, a municipality in the state of Minas Gerais, the author presented a list of shrub and tree species frequently found in the region and used as RAD measures. Figure 08 (A and B) shows the areas before and after the environmental recovery and floristic management actions. It is worth noting that figure B reflects the development of the green re-establishment of the areas after 18 years of environmental control and management.

Figure 08- Topographical reconstruction carried out in the mining areas in Poços de Caldas - MG

Source: Nappo, 2002. Available at < https://goo.gl/P9fNG6 >. Accessed on 10/10/2017.

Figure 09 shows a successful case of an area recovered by combining grass and tree species according to the concepts and practices of plant succession in the city of São Paulo, SP.

Figure 09- Recovering a degraded area by planting trees

Source: Sanchez, 2005. Available at < http://www.scielo.br/pdf/rarv/v29n1/24234>. Accessed on: 03/04/2017.

In the study carried out by Araújo, et al (2007), the authors describe the phytosociological structure of the colonising shrub vegetation in an area degraded by mining in the municipality of Brás Pires. The plant species found are shown in Table 2, which, according to the authors, provides information on the stage of plant recomposition, according to the ecological succession process (pioneer, secondary and late secondary species) to which each species is normally inserted into the system.

Tree species have numerous advantages in a reclamation process in terms of improving soil quality. The extensive root system allows trees not only to accumulate nutrients that are removed from a large volume of soil, but also to redistribute them, improving the fertility of subsurface horizons. Trees also improve soil conditioning by increasing organic matter, as well as favouring improved soil and surface microclimatic conditions. Trees create favourable conditions for mesofauna, microfauna and microflora, improving physical, chemical and biological characteristics in the rhizosphere, resulting in a profound beneficial effect on plant growth and soil characteristics (FISHER, 1995).

According to Grazziotti et al (1998), revegetation of degraded areas using tree species is more advantageous than with herbaceous species, because the former have a great capacity to immobilise larger quantities of metals and because they are more suited to stressful situations.

According to Haynes and Beare (1996), roots play an important role in the formation of aggregates. Their growth through the soil pores forces the particles closer together, which, together with the exudation of organic compounds, promotes aggregation. Another advantage of using tree species to recover degraded areas is that they provide shelter, shade and forage for wildlife (DAGANG; NAIR, 2003).

The use of the legume bragatinga (Mimosa scabrella) and the eucalyptus tree (Eucalyptus saligna) to correct the acidity and fertility of soil altered by mining is common in the Iron Quadrangle municipalities. According to Carpanezzi et al (2007), the pau-pólvora (Trema micraniha) and pau brasil (Caesalpinia echinata) trees are of great importance for use in the recovery of areas undergoing rehabilitation because they have rapid initial growth and contribute significantly to stabilising the soil's organic matter.

Table 02 - Tree species and level of plant succession used in the recovery of degraded areas.

Species	**Family**	**Successional Characteristic**
Xylopia sericea St Hill.	*Annonaceae*	Pioneer
Casearia sylvestris Swartz	*Flacourtiaceae*	Pioneer
Machaerium nyctitans (Vell.) Benth.	*Fabaceae*	Pioneer
Dalbergia nigra (Vell.) Allem.	*Leguminosae / Fabaceae*	Initial secondary
Xylopia brasiliensis Sprengel	*Annonaceae*	Initial secondary
Cecropia hololeuca Snethlage	*Cecropiaceae*	Pioneer
Cecropia glaziovii Miq.	*Cecropiaceae*	Pioneer
Anadenanthera peregrina	*Fabaceae*	Initial secondary
Platypodium elegans Vog	*Fabaceae*	Initial secondary
Matayba eleagnoides Radlk.	*Sapindaceae*	Initial secondary
Annona cacans Warm.	*Annonaceae*	Initial secondary
Tibouchina granulosa	*Melastomataceae*	Pioneer
Sparattosperma leucanthum	*Bignoniaceae*	Initial secondary
Miconia cinnamomifolia	*Melastomataceae*	Initial secondary
Lacistema pubescens Mart.	*Lacistemataceae*	Initial secondary
Rollinia sylvatica Mart.	*Annonaceae*	Initial secondary
Piptadenia gonoacantha Mart.	*Fabaceae*	Initial secondary
Solanum cernuum Vell.	*Solanaceae*	Pioneer
Hyeronima alchorneoides	*Phyllanthaceae*	Initial secondary
Schinus terebinthifolius Raddi	*Anacardiaceae*	Pioneer
Siparuna guianensis Aublet	*Siparunaceae*	Late secondary
Persea sp.	*Lauraceae*	No classification

Sapium glandulatum	*Euphorbiaceae*	Pioneer
Pseudopiptadenia contorta	*Fabaceae / Mimosoideae*	Initial secondary
Machaerium brasiliensis	*Fabaceae*	Initial secondary
Machaerium stipitatum	*Fabaceae*	Initial secondary
Vernonia diffusa Less	*Asteraceae*	Initial secondary
Dalbergia variabilis Vogel	*Fabaceae*	No classification
Miconia albicans (Sw.) Triana	*Melastomataceae*	Pioneer
Zanthoxylum rhoifolium Lam.	*Rutaceae*	Initial secondary
Nectandra opositifolia Ness	*Lauraceae*	Initial secondary
Caesalpinia echinata	*Fabaceae*	Initial secondary
Caesalpinea ferrea	*Fabaceae*	Initial secondary
Coutarea hexandra	*Rubiaceae*	Initial secondary
Piptocarpha macropoda Baker	*Asteraceae*	Pioneer
Erthtroxylum pelleterianum St. Hill.	*Erythroxylaceae*	Initial secondary
Solanum swartzianum	*Solanaceae*	Pioneer
Syagrus romanzoffiana	*Arecacea*	Initial secondary
Senna macranthera	*Fabaceae Caesalpinioidae*	Initial secondary
Cedrela fissilis Vell.	*Meliaceae*	Initial secondary
Luehea grandiflora Mart.	*Malvaceae*	Initial secondary
Jacaranda micrantha Cham.	*Bignoniaceae*	Initial secondary
Copaifera langsdorffii Desf.	*Caesalpinioideae*	Late secondary
Manihot pilosa Pohl.	*Euphorbiaceae*	Pioneer
Ficus guaranítica Chodat	*Moraceae*	No classification
Jacaranda puberula Cham.	*Bignoniaceae*	Initial secondary
Terminalia brasiliensis	*Combretaceae*	No classification
Trema micranta Blume	*Cannabaceae*	Pioneer
Ocotea cf diospyrifolia	*Lauraceae*	Initial secondary
Senna multijuga	*Caesalpinioideae*	No classification
Eucalyptus	*Eucalyptus sp.*	Initial secondary
Persea pyrifolia Ness	*Lauraceae*	Initial secondary
Byrsonima verbascifolia	*Malpighiaceae*	No classification
Tabebuia caraiba Bureau	*Bignoniaceae*	Late secondary

Source: Adapted from Sguizzatto Araújo, *et al, 2007.*

CHAPTER 5

CONCLUSION

This study concludes that although companies operating in the mining market are exposed to the need to restore degraded areas after their activities, little is publicised about the success of the practice and the techniques used to restore vegetation, a fact confirmed by the scarcity of publications on the subject.

Various factors in the process of recovery with the use of flora influence the success or decline of recovery. Climate, relief, soil structure, chemical composition, lighting and water availability are some of the factors that can interfere with the dynamics of recovery.

It is therefore necessary, before applying the species directly to the soil, to carry out studies of the regions in order to correct any non-conformities beforehand (such as the high or low presence of metals or nutrients), as well as the fact that the study is fundamental in making decisions about the techniques (soil transposition, seed bank, twig transposition, natural and artificial perches, nucleation and planting of seedlings) and species to be used.

From the analyses of the articles consulted, the most common species found were: guandu bean, bragatinga, eucalyptus, grass, brachiaria, pau brasil, arnica mineira and skunk grass. The most cited families were Fabaceae, Asteraceae, Melastomataceae and Orchidaceae, which shows a low diversity of species in the revegetation process in mining areas. In addition, the authors emphasise the high use of exotic species from Brazilian biomes, justifying their use due to their high adaptability to soil characteristics after mineral exploitation.

REFERENCES

ALVES, André Naves. **History and importance of mining in the state**. Revista do legislativo, Belo Horizonte: Legislative Assembly of the State of Minas Gerais, n. 41, p. 27-32, Jan./Dec. 2008.

BRAZILIAN ASSOCIATION OF TECHNICAL STANDARDS (ABNT). **Soil Degradation.** Available at <http://www.ebah.com.br/content/ABAAAfHu4AJ/nbr-10703- 1989-degradacao-solo>. Accessed on 07 May 2017a.

______. **Preparation and presentation of a rehabilitation project for areas degraded by mining**. Available at <http://www.ebah.com.br/content/ABAAAArUUAF/nbr-13030>. Accessed on 07 May 2017b.

BALISTIERI, PRMN; AUMOND, Juarês José. Environmental recovery in a clay mine, Doutor Pedrinho-SC. **National Symposium on the Recovery of Degraded Areas**, v. 3, p. 42-51, 1997.

BARRETO, Maria Laura. "**Mining and sustainable development**: challenges for Brazil." (2001).

BRANDI, I. V. **Study of the effectiveness of degraded area recovery plans (PRADs) for iron ore mining activities in the Iron Quadrangle region of Minas Gerais.** São Carlos School of

Engineering (USP): 1994. 114p. (Master's dissertation).

BITAR, Omar Yazbek. **Evaluation of the recovery of areas degraded by mining in the metropolitan region of São Paulo**. 1997. Doctoral thesis. University of São Paulo.

CAIRNS JR., J. Restoration, reclamation and regeneration of degraded or destroyed ecosystems = Restoration, reclamation and regeneration of degraded or destroyed ecosystems. In: SOULÉ, M.E., org. Conservation biology. Sunderland: Sinauer, 1986. p. 465-484.

CARPANEZZI, A. A. et al. Multiple functions of forests: conservation and recovery of the environment. In: **Brazilian Forestry Congress**. 1990. p. 216221.

CLEMENTS, Frederic Edward. Plant succession: an analysis of the development of vegetation. No. 242. Carnegie Institution of Washington, 1916.

COSTA, O. V. et al. Vegetative soil cover and degradation of pasture in areas of Mollisols in the south of Bahia state, Brazil. Revista Brasileira de Ciência do Solo, v. 24, n. 4, p. 843-856, 2000.

DOMENICI, Thiago. **Abandoned mines threaten communities and the environment**, 2016. Available at:<http://apublica.org/2016/03/minas-abandonadas-ameacam- communities-and-environment/> Accessed on 30 March 2017.

DOWN, C.G., STOCKS, J. Environmental impact of mining. New York: John Wiley, 1977. p. 371.

ENGEL, Vera L., and John A. Parrotta. "**Defining ecological restoration: global trends and perspectives.**" (2003). Available at: < https://www.fs.fed.us/research/publications/misc/78177-2003-Engel-Parrotta-Brazil- proceedings.pdf> Accessed on 05 May 2017.

EMBRAPA - Brazilian Agricultural Research Corporation. National Soil Survey and Conservation Service/Rio de Janeiro. **Soil conservation practices**. Rio de Janeiro, 1980. 85p. (EMBRAPA-SNLCS. Série Miscelínea).

FONSECA, João José Saraiva. **Scientific Research Methodology**. 2002.

FOSCHINI, R. C.; Ribeiro, C. A. G.; Salvador, N. B. "**Environmental legislation on the recovery of areas degraded by mineral exploitation and the use of the bond mechanism.**" Proceedings of the IV AUGM Environment Congress, São Carlos. 2009.

FRANCO, Avílio A.; DE RESENDE, Alexander Silva; CAMPELLO, Eduardo FC. **The importance of leguminous trees in the recovery of degraded areas and the sustainability of agroforestry systems.** Embrapa Caprinos e Ovinos - Article in conference proceedings (ALICE). Campo Grande, 2003.

GASPARINO, Dálgima et al. Quantification of the seed bank under different land uses in a riparian area. 2006

GOMES, Keli Cristina de Oliveira et al. **Growth of garapa seedlings in response to liming and phosphorus**. 2008.

JACOBI, Claudia Maria; FONSECA DO CARMO, Flávio; CASTRO VINCENT, Regina de. **Phytosociological study of a plant community on canga as a subsidy for the rehabilitation of mined areas in the Iron Quadrangle**, MG. Revista Árvore, v. 32, n. 2, 2008.

JESUS, Edilma Nunes de et *al.* **Natural regeneration of plant species in revegetated quarries.** 2016.

KOPEZINSKI, Isaac. **Mining x the environment: legal considerations, main environmental impacts and their modifying processes.** Editora da Universidade, Federal University of Rio Grande do Sul, 2000.

LACERDA, R., LIRA-FILHO, J. D. ; SANTOS, R. D. (2011). **Indication of tree species for urban afforestation in the semi-arid region of Paraiba.** Revista da Sociedade Brasileira de Arborização Urbana, Piracicaba, 6(1), 51-68.

LIMA, H. M., FLORES, J. C. C.; COSTA, F. L. "**Reclamation plan for degraded areas versus mine closure plan: a comparative study.**" Rem: Revista Escola de Minas 59.4 (2006): 397-402.

LOPES, Sérgio de Faria et al. Seeds dispersal of uruvalheira (Platypodium elegans vog.)(Fabaceae) in cerradão, Uberlandia, MG. **Revista Árvore**, v. 34, n. 5, p. 807813, 2010.

LORENZI, Harri. "**Árvores brasileiras:** manual de identificação e cultivo de plantas arbóreas nativas do Brasil." Nova Odessa: Editora Plantarum 352p.-col. Ed. 2 (P 210) (2002).

MARTINS, Sebastião Venâncio et al. **Seed bank as an indicator of restoration of an area degraded by kaolin mining in Brás Pires, MG**. Revista Árvore, v. 32, n. 6, p. 1081-1088, 2008.

MASCHIO, L.M.A. **Evolution, stage and characterisation of research into the recovery of degraded areas in Brazil**. In Simpósio Nacional Sobre Recuperação De Áreas Degradadas, 1, 1992, Curitiba. Proceedings... Curitiba: Fupef, 1992. p. 17-33.

MENDES, M. S. **Evaluation of ecosystem services in restoration projects in the Atlantic Forest**: consequences for adaptation to climate change - sustainability indicators. Available at: < http://www.puc- rio.br/pibic/relatorio_resumo2015/resumos_pdf/ccs/GEO/GEO-3462_Maiara%20Santos%20Mendes.pdf>. Accessed on 04/05/2017.

MEIRA-JUNIOR, M. S et al. Potential species **for recovering** areas **of** semideciduous **forest in iron exploration in the serra do espinhaço.** Bioscience Journal, 2014.p.31(1).

MMA- Ministry of the Environment. **Recovery of Degraded Areas.**
Available at <http://www.mma.gov.br/informma/item/8705- recupera%C3%A7%C3%A3o-de-%C3%A1reas-degradadas> Accessed on 01 May 2017.

MOURA, De Dalvino Jose. **Recovering Areas Degraded by Mining**. State University of Goiás, Niquelândia Unit 2015.

NAPPO, Mauro Eloi et al. **Dynamics of natural regeneration of tree and shrub species in the understory of a stand of Mimosa scabrella Bentham, in a mined area**. Poços de Caldas-MG. 2002.

NOGUEIRA, Natiélia Oliveira et *al.* **Use of legumes to recover degraded areas.** Enciclopédia Biosfera, v. 8, n. 14, p. 2012-2031,2012.

PEREIRA, Israel Marinho, et al. "**Ecological characterisation of tree species occurring in riparian forest environments, as a subsidy for the restoration of altered areas in the headwaters of the Rio Grande, Minas Gerais, Brazil.**" Ciência Florestal 20.2 (2010): 235-253.

REGENSBURGER, Brigite et *al.* **Recovery of areas degraded by clay mining through topographical regularisation, the addition of inputs and litter, and fauna attractors**. 2004.

REIS, Ademir; TRES, Deisy Regina; BECHARA, Fernando Campanhã. Nucleation **as a new paradigm in ecological restoration:"space for the unpredictable". Symposium on the recovery of degraded areas with an emphasis on riparian forests. IB: São Paulo**, p. 104-121, 2006.

RODRIGUES, R. R. et al. Restoration of tropical forests: subsidies for a methodological definition and evaluation and monitoring indicators. **Recovery of degraded areas. Viçosa: UFV**, p. 203-215, 1998.

SANCHEZ, Luis Enrique. **De-engineering: environmental liabilities in the decommissioning of industrial enterprises**. Edusp, 2001.

SGUIZZATTO DE ARAÚJO, Fernanda et al. **Structure of the shrub and tree vegetation colonising an area degraded by kaolin mining**, Brás Pires, MG. Available at < http://www.redalyc.org/html/488/48830113/>. Accessed on 12/09/2017.

SILVA, João Paulo Souza. **Environmental impacts caused by mining.** Revista espaço da Sophia 8 (2007): 1-13.

SOARES, Sílvia Maria Pereira. Restoration techniques for degraded areas. **Defence of the Postgraduate Programme in Ecology Applied to the Management and Conservation of Natural Resources. Federal University of Juiz de Fora. Juiz de Fora-MG**, p. 3-7, 2009.

SOUSA, M. M. M., et al. **Influence of Vegetation Cover on Water and Soil Losses in Erosion Plots in the Brazilian Semi-Arid.**REVISTA GEONORTE 7.26 (2016): 160-171.

SUSLICK, S. B.; OBATA, O. R.; SINTONI, A. "**Mining in the State of São Paulo:** current situation, prospects and challenges for the utilisation of mineral resources." Geosciences= Geosciences 27.2 (2008): 171-192.

SOUZA, Maurício Novaes. "**Environmental degradation and recovery and sustainable development.**" (2004) Available at <http://www.locus.ufv.br/handle/123456789/9327>. Accessed on 04/05/2017.

TRES, Deisy Regina et al. Artificial perches and soil transposition for nucleator restoration in riparian areas. **Revista Brasileira de Biociências**, v. 5, n. S1, p. pg. 312-314, 2007.

TRIVINOS, A. N. S. **Introdução à pesquisa em ciências sociais**: a pesquisa qualitativa em educação. São Paulo: Atlas, 1987.

VIEIRA, N. K.; REIS, A. O papel do banco de sementes na restauração de áreas degradadas. **NATIONAL SEMINAR**, 2003.

WILLIANS, D.D. **Rehabilitation of exhausted bauxite mines in Poços de Caldas, MG**. In: Brazilian Symposium on Exploratory Techniques Applied to Geology, 3, 1984, Salvador. Proceedings. Salvador: SBG-BA, 1984. p.464-467.

WILLIANS, D.D.; BUGIN, A; REIS, J.L.B.(Coords.). **Manual for the recovery of areas degraded by mining: revegetation techniques**. Brasília: Ibama, 1990. p.96.

Printed by Books on Demand GmbH, Norderstedt / Germany